Center for Hydrogen Safety Americas Conference 2024

AA002668

Las Vegas, Nevada, USA
21 - 23 May 2024

ISBN: 979-8-3313-0081-4

Center for Hydrogen Safety Americas Conference 2024

Las Vegas, Nevada, USA
21 - 23 May 2024

AIChE
The Global Home of Chemical Engineers

Printed from e-media with permission by:

Curran Associates, Inc.
57 Morehouse Lane
Red Hook, NY 12571

CURRAN ASSOCIATES INC.
proceedings.com

Some format issues inherent in the e-media version may also appear in this print version.

AIChE
120 Wall Street, FL 23
New York, NY 10005-4020

Phone: (800) 242-4363
Fax: (203) 775-5177

www.aiche.org

Additional copies of this publication are available from:

Curran Associates, Inc.
57 Morehouse Lane
Red Hook, NY 12571 USA
Phone: 845-758-0400
Fax: 845-758-2633
Email: curran@proceedings.com
Web: www.proceedings.com

TABLE OF CONTENTS

SESSION 5 - SAFETY CULTURE DISCUSSION

KEYNOTE SPEAKER - DAY 2

SESSION 6 - INCIDENTS & LESSONS LEARNED

SESSION 7 - SAFETY IN PRODUCTION & ELECTROLYZERS

SESSION 8 - SAFE STORAGE OF HYDROGEN

SESSION 9 - BLENDING APPLICATIONS

KEYNOTE SPEAKER - DAY 3

SESSION 10 - SAFETY OF HYDROGEN FUELING STATIONS

SESSION 11 - NEW APPLICATIONS AND APPLIED R&D

SESSION 12 - SAFETY STRATEGIES

Hydrogen – Success By Design

CONFERENCE

Center Hydrogen Safety Conference

YEAR

2024

PROCEEDING

2024 Center for Hydrogen Safety Americas Conference

GROUP

General Submissions

SESSION

Session 1 - Hydrogen Safety Fundamentals & Best Practices

TIME

Tuesday, May 21, 2024 - 9:45am to 10:05am

Authors

Diakow, P. - Presenter
Vilas, K., BakerRisk

In recent years we have seen rapid growth in hydrogen markets, with many more projected projects in the next decade. While some of this growth is associated with companies that have established track records of hydrogen operations and robust process safety programs, there are also new entrants to the market with varying levels of experience relative to quantifying and mitigating risks associated with hydrogen operations. This is further complicated for companies lacking formal process safety programs and governmental regulations / industrial standards struggling to keep pace with the rapidly expanding hydrogen market.
This presentation will introduce the CCPS book *Guidelines for Integrating Process Safety into Engineering Projects* and discuss how it can be applied to hydrogen related projects - even in cases when a facility and/or project is not

covered by PSM regulation. A case study will demonstrate how technical safety studies should move from conceptual screening in early design phases through to detailed modeling in late design/operations. This presentation will draw on years of hydrogen risk experience as well as experimental data from full-scale hydrogen explosion, dispersion, and jet fire testing to demonstrate the potential pitfalls of discounting a proactive approach to hazard/risk assessment and mitigation throughout the capital project lifecycle of Hydrogen Economy-related infrastructure.

Process Safety Management for the Hydrogen Industry

CONFERENCE

Center Hydrogen Safety Conference

YEAR

2024

PROCEEDING

2024 Center for Hydrogen Safety Americas Conference

GROUP

General Submissions

SESSION

Session 1 - Hydrogen Safety Fundamentals & Best Practices

TIME

Tuesday, May 21, 2024 - 10:05am to 10:20am

Authors

Moore, D. - Presenter, AcuTech Consulting Group

As the hydrogen industry builds out, it will need to maintain the highest of safety standards to justify the validity of the industry for its many applications. Process Safety Management (PSM) is a proven approach to managing risks of facilities handling highly hazardous chemicals and is the global benchmark for managing catastrophic risks of chemical-related operations. All in the hydrogen supply chain would benefit from an enterprise approach to PSM voluntarily applied even if not required by regulation, in addition to industry codes and standards.

Experience of 30 years of PSM has shown that it has changed the way business is managed and is deeply integrated into day-to-day operations. This paper will explain the application of PSM to hydrogen operations in manufacturing, production, and its use as a fuel. Lessons learned will be explained to show the value and benefits.

This paper will be particularly useful for companies who are producers or users of hydrogen, making them aware of the advantages of a formal management system for PSM. It will also benefit organizations practicing PSM looking for improved performance.

Note: This was presented at the Washington DC Hydrogen Safety Panel (HSP) meeting and it was agreed that this is a necessity to manage risks, especially in operation of hydrogen facilities. The DOE HSPâ€™s Safety Planning process may be appended to include consideration of PSM guidance in future. A committee was formed to work on this. As such, this is an important message for the audience at the CHS Americas meeting.

Hydrogen Recombiners: An Alternative Strategy for Hydrogen Mitigation

CONFERENCE

Center Hydrogen Safety Conference

YEAR

2024

PROCEEDING

2024 Center for Hydrogen Safety Americas Conference

GROUP

General Submissions

SESSION

Session 1 - Hydrogen Safety Fundamentals & Best Practices

TIME

Tuesday, May 21, 2024 - 10:20am to 10:35am

Authors

Gardner, L. - Presenter
Liang, Z.
Zanoni, M.
Sang, L.

Hydrogen recombiners exploit a noble metal catalyst that enables the oxidation of hydrogen by oxygen (in air) reaction for non-flammable hydrogen-air mixtures at room temperatures. They can be used to remove hydrogen in confined spaces during accidental release, preventing the potential for accumulation, fire, and explosions, or they can be used as part of ventilation systems in controlled releases. The most common hydrogen recombiner found in the market â€“ the passive autocatalytic recombiner (PAR) â€“ has been widely adopted in the nuclear industry to prevent hydrogen explosions in nuclear containment buildings. To-date, despite their usefulness, hydrogen recombiners are not widely known outside the nuclear industry.

More recent studies have focused on the application of hydrogen recombiners for the hydrogen industry, which can take advantage of years of research experience. A new study has been performed at Canadian Nuclear Laboratories (CNL) to compare the economics of retrofitting a hydrogen installation with an upgraded active ventilation system versus the installation of recombiners to prevent the accumulation of hydrogen. This paper will provide details on the relevant studies of recombiners, including an economic analysis, to demonstrate the applicability of hydrogen recombiners as an alternative or redundant approach for hydrogen mitigation.

Safety Approach and Accomplishments during Fuel Cell Technology Demonstration at Microsoft Data Center

CONFERENCE

Center Hydrogen Safety Conference

YEAR

2024

PROCEEDING

2024 Center for Hydrogen Safety Americas Conference

GROUP

General Submissions

SESSION

Session 1 - Hydrogen Safety Fundamentals & Best Practices

TIME

Tuesday, May 21, 2024 - 10:35am to 10:50am

Authors

Parente, S. - Presenter

Hydrogen is seen as a potential solution to the ever-growing power demands of the datacenter market which in 2022 accounted for 240 â€“ 340 TWh of global electricity consumptioni. As part of exploring hydrogenâ€™s viability as a future solution, Caterpillar Inc. in collaboration with Microsoft and Ballard Power Systems recently demonstrated the use of large-format hydrogen fuel cells to supply reliable and sustainable backup power for data centers. The demonstration provided valuable insights into the capabilities of fuel cell systems to power multi-megawatt data centers, ensuring uninterrupted power supply to meet the 99.999% uptime requirements. At the same time, the project showcased the learning curve facing the power industry around identifying and implementing the best practices that have enabled safe expansion of the hydrogen market in the industrial sector over many decades.

This presentation will focus on how Caterpillar, Microsoft, NREL, and other partners collaborated to manage this safety learning curve, address gaps in current EHS approaches, establish requirements and activities that helped drive zero (0) lost time incidents on the project. We will take a deep dive into the safety approach implemented for this project. We will show how early planning, coordination with our partners, safety professionals, and subject matter expertsâ€"including CHS Hydrogen Safety Panelâ€"facilitated our journey navigating the hazards of novel technologies such as large format fuel cells and battery energy storage. We will review our approach to identification of safety vulnerabilities and risk reduction and the mindset changes that were required to drive safety outcomes.

Hydrogen Regulations, Codes, and Standards Landscape

CONFERENCE

Center Hydrogen Safety Conference

YEAR

2024

PROCEEDING

2024 Center for Hydrogen Safety Americas Conference

GROUP

General Submissions

SESSION

Session 2 - Regulations, Codes, and Standards

TIME

Tuesday, May 21, 2024 - 11:40am to 12:00pm

Authors

Hamilton, J. - Presenter

The world of codes and standards is a dynamic one with many documents undergoing consistent revision cycles. While industrial hydrogen standards have been in place for decades, in the nascent hydrogen market of fuel cells, vehicles of all kinds, refueling infrastructure, power generation, backup power, long duration energy storage, etc. there is much activity in numerous standards development organizations. There is also a buzz in the legislative and regulatory arenas around hydrogen, and one may well feed into the other. This talk will discuss some of the recent activities in hydrogen codes and standards, including new standards under development and revisions of existing documents, reflecting a need from the industry and market. There is also much effort to reach out and educate permitting authorities about such codes and standards, and work to identify gaps in knowledge and practical application. This talk will highlight resources that are available to assist in those efforts. The hydrogen regulations, codes and standards landscape is broadening quickly to reflect the multitude of uses for hydrogen in decarbonizing our energy and transportation sectors, and this talk will highlight some of that activity.

IEA TCP Hydrogen Task 43 - Safety and Rcs of Large Scale Hydrogen Energy Applications: Progress

CONFERENCE

Center Hydrogen Safety Conference

YEAR

2024

PROCEEDING

2024 Center for Hydrogen Safety Americas Conference

GROUP

General Submissions

SESSION

Session 2 - Regulations, Codes, and Standards

TIME

Tuesday, May 21, 2024 - 12:00pm to 12:15pm

Authors

Tchouvelev, A. V. - Presenter
Vyazmina, E., AIR LIQUIDE
Shibutani, T.
Barilo, N., AIChE
Mey, N.
Chang, R.
Torrado, D.
Groth, K., University of Maryland
de Reals, G.
Phillips, L.

Emerging Large Scale Hydrogen Energy Applications create new challenges for the regulatory framework. To prevent and mitigate potential barriers for the deployment of large scale hydrogen applications, the IEA TCP Hydrogen Task 43 started to work on the following hydrogen safety topics: Social (Comprehensive) Risk; Safety Culture and Management System; Safety Distance Methodologies; Hazardous Areas Methodologies; Hydrogen System Safety and Dissemination. The main goal of the Task on Hydrogen Safety is to develop effective risk management methodologies and recommendations via case studies and data analysis, as well as targeted information products that will facilitate the accelerated market penetration of large-scale hydrogen energy systems.
This presentation gives the overview on the progress of each subtask highlighting the findings and showing the way forward for the harmonization of developed methodologies. The progress and success of Task 43 is crucial for relevant international and national RCS development activities. It is also instrumental in helping global hydrogen industry with market development and establishment of best safety practices.

Keywords: *Hydrogen Energy Applications, RCS, horizontal safety*

Beta-Release of Hydrogen Safety Codes and Standards Applicability Navigator (HySCAN) Informational Tool

CONFERENCE

Center Hydrogen Safety Conference

YEAR

2024

PROCEEDING

2024 Center for Hydrogen Safety Americas Conference

GROUP

General Submissions

SESSION

Session 2 - Regulations, Codes, and Standards

TIME

Tuesday, May 21, 2024 - 12:15pm to 12:30pm

Authors

Ehrhart, B. - Presenter
Hartmann, K.
Barilo, N., AIChE
Dalton, A.
Buttner, W.

There are new stakeholders entering into discussions about hydrogen system design and safety, and there are many existing codes and standards that relate to the design and safety of hydrogen systems. Some of these codes and standards are relevant to nearly any system, while others may refer to a very specific component within a system. An informational tool called Hydrogen

Safety Codes and Standards Applicability Navigator (HySCAN) has been released to the public as an initial â€œbetaâ€□ version. The overall goal of this tool is to make it easier for new and existing stakeholders of hydrogen systems to identify what codes and standards exist currently and may apply to their particular system. A user can list what type of components that a system of interest has, answer some short checklist-style questions, and the tool can then return a list of potentially relevant or useful hydrogen codes and standards. It should be emphasized that this is not a definitive list of codes and standards, nor should this be viewed as anything more than an informational tool. However, this tool should help identify what codes and standards exist for a particular system type, which can be informative for all stakeholders involved, even if those codes and standards are not legally required for that jurisdiction. This can help inform productive conversations between all stakeholders around a hydrogen system, including developers, authorities, and members of the public. The initial beta-version of the tool will be demonstrated, and feedback solicited.

Hydrogen Sealability: Alt. Designs, Materials, Facing

CONFERENCE

Center Hydrogen Safety Conference

YEAR

2024

PROCEEDING

2024 Center for Hydrogen Safety Americas Conference

GROUP

General Submissions

SESSION

Session 3 - Hydrogen Materials & Risk Management

TIME

Tuesday, May 21, 2024 - 1:45pm to 2:00pm

Authors

General, J. - Presenter

The hydrogen industry has been identified as a key decarbonization technology with its demand projected to grow five-fold, which can be seen by the influx of capital investment from the United States and globally. Several new projects are breaking ground with the goal of producing 50 million metric tons/year by 2050 in the United States and 500 million metric tons/year globally.

Currently, for several of the standards that are being used, such as ASME B31.12-2019 [1] (Hydrogen Piping and Pipelines), a majority of the sealing components call for standard gaskets from ASME B16.20 [2] and ASME B16.21 [3]. However, much is still relatively unknown regarding the potential leak rates of these standard gaskets or if the use poses any potential risks. In order to support the industry, we have performed EN 13555 [4] leak testing using 100% hydrogen across several different gaskets: ASME B16.20 [2] spiral wound, grooved metal gaskets, and alternate geometries and constructions, ASME B16.21 [3] rings with different materials including expanded polytetrafluoroethylene (ePTFE), nitrile butadiene rubber (NBR), and fluoroelastomer (FKM).

From what was observed, the permeability of hydrogen from the gaskets evaluated varies based on several different factors such as: sealing design, material, and facing material, some of which has a greater effect on creating a seal. This paper will discuss in depth the results of these gaskets tested with hydrogen and additionally make correlations to its sealability using helium.

The Next Evolution in Hydrogen Risk Analysis: “Knowledge Based” Process Hazards Analysis Utilising Developments in Digital Technology and Platforms

CONFERENCE

Center Hydrogen Safety Conference

YEAR

2024

PROCEEDING

2024 Center for Hydrogen Safety Americas Conference

GROUP

General Submissions

SESSION

Session 3 - Hydrogen Materials & Risk Management

TIME

Tuesday, May 21, 2024 - 2:00pm to 2:15pm

Authors

Scott, M. - Presenter, aeShield
Matiti, K.

The current â€œbest practiceâ€□ for executing Process Hazards Analysis (PHA) or Layer of Protection Analysis (LOPA) studies often results in less than desirable outcomes within an end user organization such as:

- Inconsistent Risk Ranking between like process units, e.g. electrolyzers / purification units / compressors / vents, when studies are conducted by different facilitators / study teams.
- Over estimation of Risk resulting in potentially unwarranted spend to manage / close Risk Gaps.
- Under estimation of Risk resulting in potentially higher than acceptable Risk Gaps remaining invisible to the end user.

Poor risk analysis execution especially in the new hydrogen industry sphere, has significant long-term cost impacts as well as increased business risk for end user organizations.

Adopting a â€œKnowledge Basedâ€□ study methodology in the new hydrogen industry represents the next evolution of PHA or LOPA study execution. The term â€œKnowledge basedâ€□ relates to studies that templatize recurring hazards associated with e.g. electrolyzers / purification units / compressors / vents, so that they are not missed, these are based on historical incidents and previous analysis, while maintaining the workshop aspects to allow personnel to review and add to prior historical findings. This white paper will highlight how â€œKnowledge Basedâ€□ PHA / LOPA execution within the hydrogen industry can result in:

- Reduction in study team time.
- Consistent risk ranking for like process units regardless of facilitator(s) / study team(s) personnel.
- Utilization of digital technology and platforms to ensure that prior knowledge is captured and maintained.
- Improved study quality.

PEM Electrolysis Failure Scenarios Identified By Failure Modes and Effects Analysis (FMEA)

CONFERENCE

Center Hydrogen Safety Conference

YEAR

2024

PROCEEDING

2024 Center for Hydrogen Safety Americas Conference

GROUP

General Submissions

SESSION

Session 3 - Hydrogen Materials & Risk Management

TIME

Tuesday, May 21, 2024 - 2:15pm to 2:30pm

Authors

Groth, K. - Presenter, University of Maryland
Wismer, S.
Al-Douri, A.
Jimenez, A.

PEM electrolysis is the most promising technology for clean hydrogen production. Quantitative risk assessment (QRA) is used to support safe and reliable deployment and operation of electrolyzers. A bottom-up failure modes and effects analysis (FMEA) was performed on a lab-scale PEM electrolyzer under normal operating conditions. We identified three significant hazards (hydrogen release, oxygen release, and hydrogen-oxygen mixing), and accompanying relevant scenarios, that apply to electrolyzers. There were 20 failure scenarios found that could lead to a hydrogen release, 45 for an oxygen release, and 68 for hydrogen and oxygen mixing. The ignition of either a hydrogen release or a collection of hydrogen and oxygen gas could cause a jet fire or explosion that poses risk to both facility personnel and surrounding equipment. The release of oxygen can also pose health risks to surrounding personnel and cause damage to nearby equipment over time. These scenarios are classified in terms of the component of interest, its failure mode, the sequence of events following failure, and the end consequence. The gas-water separators, pump responsible for supplying water to the electrolyzer stack, and the stack itself were found to contribute to the largest number of high-risk scenarios. This paper describes all of the scenarios. It supports the identification and mitigation of safety and reliability issues seen in current hydrogen electrolyzer technology as well as set the foundation for a complete QRA on hydrogen electrolyzers. Consequently, these findings will allow risk-informed decision-making for the design, operation, and deployment of PEM electrolyzers.

Public Facing Safety and Education for Hydrogen Fueling Infrastructure

CONFERENCE

Center Hydrogen Safety Conference

YEAR

2024

PROCEEDING

2024 Center for Hydrogen Safety Americas Conference

GROUP

General Submissions

SESSION

Session 4 - Public Education

TIME

Tuesday, May 21, 2024 - 3:15pm to 3:30pm

Authors

Grant, J. - Presenter

Building safe and convenient fuelling stations is key to deploying the arrival of commercial/public-use fuel cell electric vehicles (FCEVs). As the most public-facing hydrogen applications, second only to the FCEVs, hydrogen stations are an efficient tool to educate the public about hydrogen safety and normalize its use to fill up our vehicles. However, as an emerging technology, it is the industryâ€™s responsibility to ensure that fuelling infrastructures are designed and maintained in accordance with established safety standards and, thus, that the fuelling process is inherently safe for all users. On the other end, it is essential that consumers have all the necessary information at reach to help them feel safe while fuelling their zero-emission vehicles.

This paper will provide a snapshot of the safety systems used to help protect members of the public using hydrogen fueling stations, as well as the information used to educate people using this equipment. This will cover the different processes involved in hydrogen fueling stations, the dangers that are present to customers and members of the public at these sites, and the engineering design choices and equipment used to mitigate these dangers or prevent them from happening. Finally, this paper will discuss the crucial role of understanding the dangers of hydrogen at a public level and showing the importance of educating the public about hydrogen infrastructure so that people will feel comfortable using it in their everyday lives.

Public Awareness and First Responder Training in India on View of National Green Hydrogen Mission 2023

CONFERENCE

Center Hydrogen Safety Conference

YEAR

2024

PROCEEDING

2024 Center for Hydrogen Safety Americas Conference

GROUP

General Submissions

SESSION

Session 4 - Public Education

TIME

Tuesday, May 21, 2024 - 3:30pm to 3:45pm

Authors

Kelkar, G. - Presenter

India published its National Green Hydrogen Mission last year on 04/01/2023 with an objective â€œTo make India the Global Hub for production, usage and export of Green Hydrogen and its derivatives.â€□ Key enablers are identified in the mission statement where Public awareness is one of them.
Current demand of hydrogen in India is ~6 MMTPA (poised to reach 12MMTPA by 2030) driven by captive consumption of refineries and fertilizer units. The residual consumption is distributed among the steel and chemical industries. Use of Hydrogen is not new however, it was never a part of public usage (vehicles) and therefore public awareness is very crucial.

I have carried out a comprehensive survey in India about Hydrogen safety. I categorized the survey in general public, people from energy sector and students. Survey highlights that 64% public have not heard about Hydrogen till date and 30% think that it will be dangerous.

Survey emphasizes the need for rigid plans for public awareness and first responders training in India. Training shall include hydrogen properties, storage, FCEV, fire and fire-fighting by water, Battery hazards, PPEs, safety measures TRPDs etc.

Mode of trainings and awareness can be through internet, newspapers, advertisements, government programs, informative websites, courses for officials and students.

Hydrogen is very clean fuel and its usage will definitely solve global warming and pollution issues. It is important that the Mission becomes widely accepted across the country and is safely implemented too.

Safety Culture

CONFERENCE

Center Hydrogen Safety Conference

YEAR

2024

PROCEEDING

2024 Center for Hydrogen Safety Americas Conference

GROUP

General Submissions

SESSION

Session 5 - Safety Culture Discussion

TIME

Tuesday, May 21, 2024 - 3:55pm to 4:10pm

Access to cryogenic hydrogen training facilities has been historically limited and remains a severe impediment to Research and Development efforts. The US Inflation Recovery Act of 2022 is resulting in unprecedented investment in hydrogen technologies and exacerbating this training deficiency. To address this need, the Hydrogen Properties for Energy Research (HYPER) Center at Washington State University developed the Cool Fuel School cryogenic hydrogen safety program in the fall of 2023 with support from Airbus. This new approach teaches safety plan development integrated with the engineering design process while guiding participants through the clean-sheet design, development, and testing of their very own liquid hydrogen flow manifold. This presentation discusses the design of the Cool Fuel School, including pre-requisite education on cryogenic hydrogen theory, hands-on workshops, and lessons learned from preliminary participants.

Safety Culture – The Missing Link

CONFERENCE

Center Hydrogen Safety Conference

YEAR

2024

PROCEEDING

2024 Center for Hydrogen Safety Americas Conference

GROUP

General Submissions

SESSION

Session 5 - Safety Culture Discussion

TIME

Tuesday, May 21, 2024 - 4:10pm to 4:30pm

Authors

Bayley, S. - Presenter, Powertech USA, Inc. - LIFTE H2, Inc.

Access to cryogenic hydrogen training facilities has been historically limited and remains a severe impediment to Research and Development efforts. The US Inflation Recovery Act of 2022 is resulting in unprecedented investment in hydrogen technologies and exacerbating this training deficiency. To address this need, the Hydrogen Properties for Energy Research (HYPER) Center at Washington State University developed the Cool Fuel School cryogenic hydrogen safety program in the fall of 2023 with support from Airbus. This new approach teaches safety plan development integrated with the engineering design process while guiding participants through the clean-sheet design, development, and testing of their very own liquid hydrogen flow manifold. This presentation discusses the design of the Cool Fuel School, including pre-requisite education on cryogenic hydrogen theory, hands-on workshops, and lessons learned from preliminary participants.

Safety Culture Panel

CONFERENCE

Center Hydrogen Safety Conference

YEAR

2024

PROCEEDING

2024 Center for Hydrogen Safety Americas Conference

GROUP

General Submissions

SESSION

Session 5 - Safety Culture Discussion

TIME

Tuesday, May 21, 2024 - 4:30pm to 4:50pm

Authors

Moore, D. - Presenter, AcuTech Consulting Group

Access to cryogenic hydrogen training facilities has been historically limited and remains a severe impediment to Research and Development efforts. The US Inflation Recovery Act of 2022 is resulting in unprecedented investment in hydrogen technologies and exacerbating this training deficiency. To address this need, the Hydrogen Properties for Energy Research (HYPER) Center at Washington State University developed the Cool Fuel School cryogenic hydrogen safety program in the fall of 2023 with support from Airbus. This new approach teaches safety plan development integrated with the engineering design process while guiding participants through the clean-sheet design, development, and testing of their very own liquid hydrogen flow manifold. This presentation discusses the design of the Cool Fuel School, including pre-requisite education on cryogenic hydrogen theory, hands-on workshops, and lessons learned from preliminary participants.

Overview of the U.S. Department of Energy's Hydrogen and Fuel Cell Technologies Office Safety, Codes & Standards Activities

CONFERENCE

Center Hydrogen Safety Conference

YEAR

2024

PROCEEDING

2024 Center for Hydrogen Safety Americas Conference

GROUP

General Submissions

SESSION

Keynote Speaker - Day 2

TIME

Wednesday, May 22, 2024 - 9:05am to 9:45am

Authors

Hill, L. - Presenter

For over two decades, the Department of Energyâ€™s (DOEâ€™s) Hydrogen and Fuel Cell Technologies Office (HFTO) has led Research, Development, & Demonstration (RD&D) activities to enable the commercial viability and safe adoption of hydrogen and fuel cell technologies that benefit multiple applications and sectors. Given the value of hydrogen and fuel cells beyond transportation, the HFTO RD&D focus includes industrial uses, energy storage, grid integration, and heavy-duty transportation applications.

To enable widespread commercialization and safe deployment of these emerging applications, HFTOâ€™s Safety, Codes, & Standards (SCS) activities provide the fundamental understanding of the relevant physics, critical data, and safety information used to develop and revise technically sound and defensible codes and standards. SCS identifies and evaluates safety and risk assessment approaches that can be used to define requirements for and close gaps in codes and standards in a timely manner. SCS collaborations among researchers in government, industry, and academia focus on advanced analysis of critical hydrogen behavior data and development of safety and sensing technologies to support the development and deployment of hydrogen codes and standards. These collaborations also identify tools and methodologies for dissemination through best practices and lessons learned, such as on the H2Tools.org site and through the Hydrogen Safety Panel.

Lessons Learned from a Medium Pressure Tank Leak at Nrel's Hydrogen Refueling Station

CONFERENCE

Center Hydrogen Safety Conference

YEAR

2024

PROCEEDING

2024 Center for Hydrogen Safety Americas Conference

GROUP

General Submissions

SESSION

Session 6 - Incidents & Lessons Learned

TIME

Wednesday, May 22, 2024 - 10:00am to 10:15am

Authors

Clark, P. - Presenter
Hartmann, K.
Smith, O.
Leighton, D.

Safe and reliable hydrogen systems are critical to the expansion and widespread utilization of hydrogen as an energy storage mechanism. Hydrogen component failures leading to external leaks create a hazard which could lead to fire or explosion. Leaks with the most potential energy are those which cannot be isolated from the bulk hydrogen storage system. In 2021 a leak occurred on a union connected to the end plug of a medium pressure tank at the National Renewable Energy Laboratoryâ€™s Hydrogen Fueling Station. The failure of an O-ring in the union created an external leak which could not be isolated. There was no ignition or any injuries as a result of this leak event. Here we outline the potential cause of the leak, the lessons learned, and the improvements to the hardware design. NRELâ€™s highly instrumented station provides a clear indication of the leak with pressure, temperature monitoring, and ultrasonic leak detection instruments. Post failure analysis provides insight as to why the O-ring failed and informed the redesign of the medium pressure gas manifold. This leak occurred during the recommissioning of the medium pressure system following extended downtime and rebuild of a system that had been originally installed at a different facility. Sharing this experience informs the greater hydrogen community and hopefully can be utilized to inform and improve other hydrogen systems.

Hydrogen Plant Relief Line Failure Due to Acoustically Induced Vibrations

CONFERENCE

Center Hydrogen Safety Conference

YEAR

2024

PROCEEDING

2024 Center for Hydrogen Safety Americas Conference

GROUP

General Submissions

SESSION

Session 6 - Incidents & Lessons Learned

TIME

Wednesday, May 22, 2024 - 10:15am to 10:30am

Authors

Wohlgeschaffen, K. - Presenter
Galloway, G.

A relief line section of piping failed unexpectedly during a startup of one of our hydrogen plants. Fortunately no one was in the area at the time of release and the escaping syngas did not ignite or explode. To shorten startup duration, this startup line was utilized to vent the heat-up gas to relief downstream of the low temperature shift converter reactor. The line failure was attributed to fatigue caused by a lesser known damage mechanism, acoustic induced vibration ("AIV"). AIV occurred due to the line not being designed for the full gas flowrate. This incident highlights the importance of having and following well

written procedures, recognizing, and mitigating risk, and completing thorough risk assessments through the Management of Change process. It also serves to raise awareness around AIV, and is a good reminder of the industry need to avoid the common operational trap of normalizing deviation.

CGA H-17, Guideline for Small Scale Hydrogen Production

CONFERENCE

Center Hydrogen Safety Conference

YEAR

2024

PROCEEDING

2024 Center for Hydrogen Safety Americas Conference

GROUP

General Submissions

SESSION

Session 7 - Safety in Production & Electrolyzers

TIME

Wednesday, May 22, 2024 - 11:10am to 11:30am

Authors

Early, R. - Presenter
Early, R. - Presenter

With the increased demand for hydrogen for fuel cell electric vehicles (FCEVs), smaller hydrogen production facilities are becoming more common. Unlike industrial sites for hydrogen production, these smaller plants and the loading facilities associated with them can be more visible and accessible to the public. The end users of this hydrogen might have little or no experience in the chemical process industry. In addition, these facilities are more likely than large hydrogen plants to be operated unattended or partially attended (not around the clock).

CGA H-17, *Guideline for Small Scale Hydrogen Production*, provides a single comprehensive set of safety and operating guidelines to address the multiple hazards associated with the operation of small-scale hydrogen production plants that use various hydrogen production methodologies.

Some of the main issues that impact small scale hydrogen production facilities and the end users are listed below:

- Mitigating process hazards from flammable, toxic, and high temperature gases, high voltage electricity, and hazardous process materials such as water treatment chemicals, catalysts, and adsorbents.
- Dealing with hazards from a bulk liquid hydrogen backup storage system, which may be large enough to require a plan to comply with OSHA Process Safety Management (PSM) and EPA Risk Management Program (RMP) regulations.
- Minimizing the impact of co-locating the production facility with a vehicle fueling station and dealing with public users of the fueling station.
- Mitigating the safety concerns associated with a partially staffed (lone worker) or unstaffed production facility.

Electrical Area Classification Safety-in-Design Considerations for a Large Alkaline Electrolyzer Hydrogen Plant

CONFERENCE

Center Hydrogen Safety Conference

YEAR

2024

PROCEEDING

2024 Center for Hydrogen Safety Americas Conference

GROUP

General Submissions

SESSION

Session 7 - Safety in Production & Electrolyzers

TIME

Wednesday, May 22, 2024 - 11:30am to 11:50am

Authors

Gardner, T. - Presenter

Introduction

This abstract focus is safety in design considerations relative alkaline electrolysis equipment (pressurized) electrical area classification as applies to large hydrogen production facilities. Alkaline electrolyzers are unique due to size/capacity, and open style design. While alkaline electrolyzer technology is quite mature, we explore common design challenges around electrical area classification.

The basis of this abstract was formed in large part from Lessons Learned on the ACES project, the worldâ€™s largest industrial green hydrogen facility and storage project in Delta, Utah, for Advanced Clean Energy Storage I, LLC (â€œACESâ€□).

Hydrogen Safety Design Basis Considerations

Pressurized Alkaline electrolyzers with open style design, with no individual enclosure around the stack, and located indoors will be discussed. Electrolyzer building ventilation considerations, and the overall space of the room containing the electrolyzer and other hydrogen generation equipment, as well as hazardous areas classifications per NEC, section 500 and 501, Class 1 Division 2, Group B will be examined. Considerations and challenges related to rating the space around the electrolyzer stack, including the DC electrical connections, areas immediately surrounding the electrolyzer, and areas to protect personnel or other objects that could cause an arc are part of safety design discussion, as well as leak sources, ventilation equipment interlocks, hydrogen leak detection instruments, LEL/LFF limit calculations, manufacturers recommendations for ventilation / air changes, location and number of hydrogen leak detectors, and design of both passive and active ventilation.

Finally, this abstract will also consider prescriptive design analysis using modeling, vs. a prescriptive code driven approach. Many international standards allow use of non-classified equipment, such as IEC 60079-13 and ISO 2274, while US codes and standards appear to be more prescriptive.

Codes and Standards

IEC 60079-13, ISO 22734, NFPA 2, NFPA 70, Section 500c, 501, NFPA 497

Large-Scale Electrolysis Plant Design and Emerging Safety Considerations, an EPC's Perspective

CONFERENCE

Center Hydrogen Safety Conference

YEAR

2024

PROCEEDING

2024 Center for Hydrogen Safety Americas Conference

GROUP

General Submissions

SESSION

Session 7 - Safety in Production & Electrolyzers

TIME

Wednesday, May 22, 2024 - 11:50am to 12:05pm

Authors

Distler, J. - Presenter

Black & Veatch is currently designing and constructing large-scale electrolysis plants throughout the U.S. Building first-of-a-kind electrolysis facilities at scale requires the proper balance of minimizing costs while maximizing safety. Gaps and inconsistencies in codes such as NFPA 2 and CGA require EPCs to perform additional due diligence when incorporating safety by design principles. This talk will discuss Black & Veatchâ€™s experience with safety by design in its projects, including equipment spacing requirements, hydrogen detection considerations, venting vs. flaring, the production of ammonia as an energy carrier with additional safety considerations, implications of hourly load following on electrolyzer selection and design, and other subjects. This presentation will also describe best practices when discussing hydrogen and associated safety requirements with clients and the public to minimize misinformation about hydrogen. The presentation aims to provide a starting point for discussion between all stakeholders in the hydrogen economy to understand how to best design and implement safety by design in all large-scale electrolysis projects. The safe production of large-scale hydrogen is imperative to increase public confidence and build industry trust, both of which are required for scaling these projects.

Facility Siting and Risk Analysis of a PEM Electrolyzer

CONFERENCE

Center Hydrogen Safety Conference

YEAR

2024

PROCEEDING

2024 Center for Hydrogen Safety Americas Conference

GROUP

General Submissions

SESSION

Session 7 - Safety in Production & Electrolyzers

TIME

Wednesday, May 22, 2024 - 12:05pm to 12:20pm

Authors

Armstrong, C. - Presenter
Ruiz-Tagle, A., AcuTech Group, Inc.
MacNguyen, R.

Presenting our current work on the QRA study of a commercial PEM electrolyzer for hydrogen production, this paper is tailored to address the unique challenges posed by hydrogen releases in production units. Recognizing that various factors can influence QRA outcomes, the study includes considerations such as incorporating the latest release frequency data for hydrogen applications, determining realistic consequences based on concentration criteria, estimating the probability of ignition, and accounting for sensor-based detection and isolation, among other factors.

IEA TCP Task 43 - Subtask Safety Distances: Electrolyser

CONFERENCE

Center Hydrogen Safety Conference

YEAR

2024

PROCEEDING

2024 Center for Hydrogen Safety Americas Conference

GROUP

General Submissions

SESSION

Session 7 - Safety in Production & Electrolyzers

TIME

Wednesday, May 22, 2024 - 12:20pm to 12:35pm

Authors

Tchouvelev, A. V. - Presenter
Vyazmina, E., AIR LIQUIDE
Chang, R.
Sébastien, Q.
Truchot, T.
Hocquet, J., Technip Energies
Torrado, D.
Hart, N.
Runefors, M.
Ehrhart, B.
Jordan, T.
Ramsey-Idem, K.
Halinen, M.

Bouchet, L.
Pérez, R. A.
de Reals, G.
Phillips, L.
Houssin, D.
Jallais, S., AIR LIQUIDE

Today local regulations for hydrogen installations differ depending on the country. They provide different safety and/or separation distances for the same hydrogen installation. The purpose of this work is to give an insight on different methodologies and recommendations developed for risk management and consequences assessment of accidental scenarios. This is achieved for the case study of an industrial electrolyser (alkaline and PEM technologies). The work includes the lessons learned from previous incidents including recommendations for the mitigation barriers, a review of the available methodologies, and identification of the divergent points. Moreover, it makes suggestions for the harmonization approach. This work also sheds light on the most impacting scenarios and the associated leakage size.

The Importance of Integrated Safety Applied to Large Scale Gaseous Hydrogen Storage

CONFERENCE

Center Hydrogen Safety Conference

YEAR

2024

PROCEEDING

2024 Center for Hydrogen Safety Americas Conference

GROUP

General Submissions

SESSION

Session 8 - Safe Storage of Hydrogen

TIME

Wednesday, May 22, 2024 - 2:00pm to 2:15pm

Authors

Kreutzer, C. - Presenter

The scale of hydrogen systems often increases both their complexity and the number of stakeholders involved in their construction, commissioning, and operation. The increased scale can also amplify the consequences of safety incidents that may occur. Multiple suppliers, owners, operators, and regulators contribute diverse risk tolerances, safety histories, and objectives to the project. In addition, their cost-driven motivations may conflict with stringent safety practices. Therefore, implementing a comprehensive safety strategy is crucial for large-scale projects, ensuring it is overseen by a single expert entity responsible for safety design, specifications, integration, and verification. This presentation will underscore the significance of integrated safety, exemplified by a case study of Braindrip's large-scale (10,000 - 500,000 kg+) gaseous hydrogen storage system. We will highlight key hydrogen failure scenarios, examine the influence of various upstream and downstream components and functional requirements on the system's safety, and assess hazard mitigation strategies for these scenarios. Special emphasis will be placed on the role of robust engineering controls and the challenges in mitigating risks associated with human involvement in operational and maintenance activities.

Value Proposition of Commercially Available Metal Hydrides; In Terms of Safety, Footprint and Integration to Various Hydrogen Use-Cases

CONFERENCE

Center Hydrogen Safety Conference

YEAR

2024

PROCEEDING

2024 Center for Hydrogen Safety Americas Conference

GROUP

General Submissions

SESSION

Session 8 - Safe Storage of Hydrogen

TIME

Wednesday, May 22, 2024 - 2:15pm to 2:30pm

Authors

Petrecky, J. - Presenter

Metal Hydrides are starting to gain commercial traction in stationary storage and specialized mobile applications, due to their unique value proposition, primarily around safety. This presentation provides insight into the following areas;

- Background on Metal Hydrides
- Commercial readiness
- Safety implications

- Footprint implications
- Use-cases that can benefit from metal hydrides

Gap Assessment of Natural Gas Pipeline for Hydrogen-Natural Gas Blend Transport

CONFERENCE

Center Hydrogen Safety Conference

YEAR

2024

PROCEEDING

2024 Center for Hydrogen Safety Americas Conference

GROUP

General Submissions

SESSION

Session 8 - Safe Storage of Hydrogen

TIME

Wednesday, May 22, 2024 - 2:30pm to 2:45pm

Authors

Pierce, B. - Presenter

The United States is currently undergoing a significant transition within its energy system. Today, there are about 3 million miles of natural gas pipelines in the United States. One way to potentially jump-start the energy transition has been the conversion of natural gas infrastructure to natural gas blended with hydrogen or pure hydrogen systems. In late 2023, ENTRUST Solutions Group was approached by an interstate natural gas transmission company to evaluate over 300 miles of their system for future hydrogen compatibility and continued, safe operation of the existing pipeline assets. Using existing code requirements, safety and operational codes and standards, legacy company documents, and engineering insight an evaluation of the potential for future conversion of service was developed. This presentation will discuss how we partnered with the client to work through how a future conversion of service could be performed while maintaining confidence in safety and reliability of the gas system. We will broadly discuss several key future conversion challenges and potential future mitigations.

Beyond the Hype - Mastering Safety in Hydrogen Transport and Storage

CONFERENCE

Center Hydrogen Safety Conference

YEAR

2024

PROCEEDING

2024 Center for Hydrogen Safety Americas Conference

GROUP

General Submissions

SESSION

Session 8 - Safe Storage of Hydrogen

TIME

Wednesday, May 22, 2024 - 2:45pm to 3:00pm

Authors

Granger, C. - Presenter

As the adoption of hydrogen as an alternative energy and fuel source gains momentum, it is imperative for industry leaders to prioritize safety considerations when it comes to hydrogen transport and storage.
Hydrogen Storage and Transportation: This section will delve into the critical aspects of hydrogen storage tank cylinders, addressing key considerations such as design, materials, and safety features. It will provide valuable insights into the selection and maintenance of cylinders to ensure safe and efficient hydrogen transport.

Cost Considerations: Safety is not just a matter of best practices but also a smart investment. This section will explore the relationship between safety and cost in hydrogen transport, highlighting the long-term benefits of investing in reliable and secure hydrogen transportation solutions.

Hydrogen Safety Measures: Safety protocols play a pivotal role in ensuring secure hydrogen transport and storage. This segment will examine key safety measures, including leak detection, ventilation systems, and emergency response procedures, providing attendees with practical guidance on maintaining safe hydrogen operations.

High-Pressure Hydrogen Transport Trailers:

The selection of trailers for high-pressure hydrogen transport is a critical consideration. This section will shed light on key factors such as trailer design, materials, safety features, and regulatory compliance. Attendees will gain valuable insights into choosing the right trailers to mitigate risks associated with high-pressure hydrogen transport.

Conclusion:

This presentation will empower industry leaders to make informed choices regarding safety considerations in hydrogen transport and storage. By prioritizing safety, we can build a foundation of trust and confidence in the hydrogen industry.

Safe Hydrogen-Natural Gas Blending: Research, Projects, and Addressing Gaps

CONFERENCE

Center Hydrogen Safety Conference

YEAR

2024

PROCEEDING

2024 Center for Hydrogen Safety Americas Conference

GROUP

General Submissions

SESSION

Session 9 - Blending Applications

TIME

Wednesday, May 22, 2024 - 3:45pm to 4:05pm

Authors

Ruggles, T. - Presenter

Hydrogen blending into natural gas pipelines is often seen as a means to kick start progress towards reducing the carbon intensity of our fuels, while utilizing existing infrastructure, and accelerating the early growth of the Hydrogen Economy. In the United States, natural gas serves approximately 61% of the 125 million households and is supported by roughly 3 million miles of natural gas pipelines. Many of these pipelines and the end-use equipment they supply were not originally designed for handling hydrogen blends, eliciting safety concerns, such as embrittlement, gas leakage, and more. Understanding the safe limits of these existing pipelines and downstream equipment is crucial to the future success of hydrogen blending. Extensive research has been and continues to address steel embrittlement and other material compatibility issues, hydrogen leakage, and pipeline conversions. Additionally, multiple blending projects have monitored gas distribution system safety and shared safety information and the performance of customer appliances while blending. However, research gaps exist due to the novelty of hydrogen blending at scale and the enormity of existing natural gas infrastructure and equipment. For example, we are supporting a forthcoming blending project within a community with multiple 80+ year old, megawatt-scale, mixed diesel/natural gas generators. What safety protocols should be developed to operate these generators on a mixed diesel/natural gas/hydrogen blend? This talk will provide an overview of hydrogen blending safety concerns, the research and projects addressing these concerns, and present strategies being implemented to address knowledge gaps.

Hydrogen Blending – Design and Safety Considerations for Pilot Projects at Power Generation Facilities

CONFERENCE

Center Hydrogen Safety Conference

YEAR

2024

PROCEEDING

2024 Center for Hydrogen Safety Americas Conference

GROUP

General Submissions

SESSION

Session 9 - Blending Applications

TIME

Wednesday, May 22, 2024 - 4:05pm to 4:20pm

Authors

Reusser, M. - Presenter

As utilities are looking to decarbonize their power generation assets, hydrogen is emerging as a potential pathway to achieve their CO_2 reduction targets because it is a noncarbon-emitting fuel when combusted. However, due to the uncertainty of how some of these assets will operate as hydrogen is introduced into the system, many facilities are looking to implement pilot-testing programs to temporarily blend and burn hydrogen with the natural gas fuel to measure performance (output, fuel consumption, emissions, etc.) before committing to large capital investments for permanent hydrogen co-firing capabilities.

This presentation will include discussion on the lessons learned during a real hydrogen-blending pilot test a power generation facility (recip. engine) in Michigan. In addition to safety considerations, it will also focus on the key aspects of implementing the hydrogen blending/burning demonstration tests:

- Coordination with Original Equipment Manufacturers (OEMs) to determine:
    - Turbine/engine hydrogen fueling capabilities
    - If modifications are required
    - Testing duration/frequency
- Hydrogen supply
    - Liquid vs. Gaseous deliveries
    - On-site generation
- Hydrogen storage
    - Discuss minimum setback distances in *NFPA 2 â€“ Hydrogen Technologies Code* for hydrogen storage placement within the facility
- Blending system
    - Achieving proper mixing of the natural gas/hydrogen
- Balance-of-Plant evaluation
    - Safety reviews - HAZOP
    - Hazardous Area Classification of electrical equipment

Additionally, the presentation will also discuss some of the challenges and lessons-learned from executing these pilot project studies:

- Engine sensitivities to changes in fuel composition/incomplete mixing
- Large hydrogen quantities required (supply chain issues)
- Siting of new equipment within an existing facility

Safety and Design of an East Coast Utility Hydrogen Demonstration for Commercial and Industrial Customers

CONFERENCE

Center Hydrogen Safety Conference

YEAR

2024

PROCEEDING

2024 Center for Hydrogen Safety Americas Conference

GROUP

General Submissions

SESSION

Session 9 - Blending Applications

TIME

Wednesday, May 22, 2024 - 4:20pm to 4:35pm

Authors

Harris PE, T. - Presenter, University of Illinois at Chicago

In 2023, EN Engineering (ENTRUST Solutions Group) was engaged by an east coast dual fuel utility partner to design a hydrogen blending demonstration project. While several other blending pilot projects had been completed in the United States, this project was a bit different. This project aimed to demonstrate hydrogen blending as a potential solution for commercial and industrial clients for space heating needs. This project worked through numerous safety aspects of both pure hydrogen and blended hydrogen in natural gas for both utilities and end users. This talk will detail how site safety considerations, the site hazard analysis, and how identified risks were mitigated throughout the design process and modification of site operational plans.

Checkout

This paper has an Extended Abstract file available; you must purchase the conference proceedings to access it.

Safety and Environmental Lessons from Hydrogen-Natural Gas Cofiring Demonstrations

CONFERENCE

Center Hydrogen Safety Conference

YEAR

2024

PROCEEDING

2024 Center for Hydrogen Safety Americas Conference

GROUP

General Submissions

SESSION

Session 9 - Blending Applications

TIME

Wednesday, May 22, 2024 - 4:35pm to 4:50pm

Authors

Gupta, A. - Presenter

To facilitate reduction of carbon dioxide emissions in the power industry, EPRI has participated in hydrogen-natural gas co-firing demonstrations involving several different power generation systems. Safety measures, environmental outcomes, and lessons learned from three of these demonstrations are presented here. Safety-relevant features of these demonstrations included appropriate stakeholder engagement, frequent and open communication, site/facility security, personnel training, and several other controls. Procedures were created by subject matter experts, with thorough planning and emergency response strategies. System modifications, equipment compatibility, and emissions monitoring protocols are described herein. These demonstrations reveal that hydrogen-natural gas cofiring is safe under the studied conditions. EPRIâ€™s work in this area constitutes a significant step in the ongoing shift towards more sustainable power generation.

All Aboard: Hydrogen on Rail

CONFERENCE

Center Hydrogen Safety Conference

YEAR

2024

PROCEEDING

2024 Center for Hydrogen Safety Americas Conference

GROUP

General Submissions

SESSION

Session 9 - Blending Applications

TIME

Wednesday, May 22, 2024 - 4:50pm to 5:05pm

Authors

Wechsung, A. - Presenter, Exponent
Hart, R., Exponent, Inc.
Morrison, D., Exponent Inc

Hydrogen is gaining momentum as a key component to a clean energy future. As part of this push to decarbonize sectors, new applications for hydrogen transportation, storage, and usage are being developed. While electric vehicles have increased their market share at the light duty vehicle segment, emissions originating from heavy duty transportation including long-haul trucking, rail, maritime shipping, and aviation are often considered hard-to-abate and hydrogen is proposed as one promising alternative.
Sensing business opportunities, new participants are entering these markets and look to develop solutions to meet these needs. Hydrogen presents specific challenges that developers and engineers may not fully appreciate if they look to transfer experience from handling traditional, carbon-based fuels. For

example, hydrogenâ€™s combustion properties, including its broad flammable range, low minimum ignition energy, and high burning velocity, suggest that explosion mitigation is an even more important topic and may require unique design solutions.

In the US, the Federal Railroad Administration (FRA) has authority over locomotives and locomotive tenders. However, we are not aware of specific regulations issued by the FRA for hydrogen fueled locomotives. In this presentation, we first review FRA guidance issued for test programs using another alternative fuel: natural gas. We will briefly describe current operations using compressed and liquified natural gas (CNG and LNG). Then, we will discuss how to translate lessons learned for natural gas projects to mitigate the hazards specific to hydrogen. In particular, we will address fueling operation, leak detection and response, and potential incidents including derailment and external fires.

The Cool Fuel School: Development of a Hands-on Liquid Hydrogen Safety Training Program for Early-Stage Research

CONFERENCE

Center Hydrogen Safety Conference

YEAR

2024

PROCEEDING

2024 Center for Hydrogen Safety Americas Conference

GROUP

General Submissions

SESSION

Keynote Speaker - Day 3

TIME

Thursday, May 23, 2024 - 9:05am to 9:45am

Authors

Leachman, J. - Presenter, Washington State University

Access to cryogenic hydrogen training facilities has been historically limited and remains a severe impediment to Research and Development efforts. The US Inflation Recovery Act of 2022 is resulting in unprecedented investment in hydrogen technologies and exacerbating this training deficiency. To address this need, the Hydrogen Properties for Energy Research (HYPER) Center at Washington State University developed the Cool Fuel School cryogenic hydrogen safety program in the fall of 2023 with support from Airbus. This new approach teaches safety plan development integrated with the engineering design process while guiding participants through the clean-sheet design, development, and testing of their very own liquid hydrogen flow manifold. This presentation discusses the design of the Cool Fuel School, including pre-requisite education on cryogenic hydrogen theory, hands-on workshops, and lessons learned from preliminary participants.

Reliability and Safety in Technology Development: H2 Mobility Applications

CONFERENCE

Center Hydrogen Safety Conference

YEAR

2024

PROCEEDING

2024 Center for Hydrogen Safety Americas Conference

GROUP

General Submissions

SESSION

Session 10 - Safety of Hydrogen Fueling Stations

TIME

Thursday, May 23, 2024 - 9:45am to 10:05am

Authors

Hernandez Bello, J. - Presenter

Developing technology that is safe and reliable is paramount for any successful commercial application. With this end in mind Shell Hydrogen has developed a comprehensive approach to design, build, and operate Hydrogen mobility assets in such a way that safety and reliability are at the core of the concurrent engineering processes that govern the stages of their lifecycle. This has been achieved in practice by leveraging systems engineering, reliability methods, and safety practices along with the experience and expertise of Shell Hydrogen, original equipment manufacturers, and system integrators.
Safety and reliability are important performance attributes of any engineered system where human-machine interactions are present. However, they are usually approached as afterthoughts or, in some cases, unintended

consequences of the system design and development process that must be addressed and verified in subsequent design stages. In plain words, safety and reliability are often seen as constraints that add layers of complexity and extra costs to the minimum functional system of interest.

The challenges in implementing this approach are many, ranging from access to historical data on equipment and component safety and reliability performance, to lack of standardization in the industry when dealing with hydrogen related hazards. This presentation is aimed at describing the approach in more detail, some of the early successes and failures during deployment, and the continual improvement journey that lies ahead.

Empowering Safety and Awareness: Htec's Journey in Hydrogen Fueling Station Development and Public Education Initiatives.

CONFERENCE

Center Hydrogen Safety Conference

YEAR

2024

PROCEEDING

2024 Center for Hydrogen Safety Americas Conference

GROUP

General Submissions

SESSION

Session 10 - Safety of Hydrogen Fueling Stations

TIME

Thursday, May 23, 2024 - 10:05am to 10:20am

Authors

Press, D. - Presenter

HTEC opened Canadaâ€™s first hydrogen retail fueling station in June 2018. Since then, the company has opened 4 more stations in British Columbia and has about 20 stations in the planning and development stages. HTEC has accumulated valuable expertise in safe fueling station design and operation supported by a safety-first corporate culture and is now looking for ways to educate the public and normalize the use of hydrogen fuel.

HTEC proposes to provide an overview of the best practices to keep our hydrogen fueling stations safe and initiate discussions about the role of public education in ensuring safety, focusing on three key elements:

- Ensuring safety through design and engineering: Employ proper engineering controls, including materials selection and equipment design, to minimize hydrogen-related risks, and follow recognized engineering standards and guidelines in the design of hydrogen systems.
- The importance of a strong safety culture: Implement training and campaigns within the organization to educate employees and contractors, foster a culture of open communication regarding safety issues and promote accountability.
- The challenges of educating the public and stakeholders: How can the hydrogen industry collaboratively dispel misconceptions about its safety? This could involve initiatives such as coordinated awareness campaigns, community engagement, demonstration projects, and school programs.

The presentation will showcase our on-ground experience and learnings around hydrogen safety to help the industry collectively grow, but also how it involves instilling a strong safety culture for our employees and contractors.

Statistics, Lessons Learned, and Recommendations from Analysis of the First 40 Fueling Station Component Failure Events in Hycred

CONFERENCE

Center Hydrogen Safety Conference

YEAR

2024

PROCEEDING

2024 Center for Hydrogen Safety Americas Conference

GROUP

General Submissions

SESSION

Session 10 - Safety of Hydrogen Fueling Stations

TIME

Thursday, May 23, 2024 - 10:20am to 10:35am

Authors

Al-Douri, A. - Presenter
Groth, K., University of Maryland
Robinson, O.
Hartmann, K.
Saur, G.
Buttner, W.

Safety, risk, and reliability issues are vital to ensure the continuous and profitable operation of hydrogen technologies. To address this gap, UMD and

NREL have developed and validated a structure for a hydrogen component reliability database, (HyCReD) [1]. HyCReD which can be used to collect and analyze hydrogen component reliability data to be used in QRA, process safety, and reliability engineering studies. The first 40 events coded in HyCReD presented in this work were used to validate the structure. We present detailed breakdowns of the first 40 fueling station component failure events by station subsystem, functional group, component, and component failure modes. We determined that the dispensing process and bulk storage were the two subsystems with the highest number of associated events. In terms of functional groups, the dispensing process and sensing and control groups were the two in which most associated components were present. The components in which most events occurred were nozzles, hoses and valves. The dominant failure mode observed in these events was an external leak of hydrogen to the environment. The initial results show that the HyCReD taxonomy and reporting structure provide important, detailed insights into the dominant reliability concerns for hydrogen fueling stations. We are actively seeking more participants to provide data for us to code; with more events from participants, we can obtain the first hydrogen-specific component failure rates. These results will provide actionable reliability engineering-based insights to enable prioritizing maintenance activities, eliminate costly downtime, and improve future station designs.

Overview of Doe Activities Addressing Hydrogen Release Concerns

CONFERENCE

Center Hydrogen Safety Conference

YEAR

2024

PROCEEDING

2024 Center for Hydrogen Safety Americas Conference

GROUP

General Submissions

SESSION

Session 11 - New Applications and Applied R&D

TIME

Thursday, May 23, 2024 - 11:20am to 11:40am

Authors

Watson, C. - Presenter
Saur, G.
Hartmann, K.
Groth, K., University of Maryland
Buttner, W.

Industry stakeholders, researchers, and many among the hydrogen safety community have recently been asked to respond to public concerns that indirect impacts of hydrogen releases on global warming can reduce the projected decarbonization benefits of hydrogen. Major topics of concern include the need for improved monitoring of potential releases across large-scale projects, the need for new knowledge and more precise data about

hydrogen release rates, and objective information about the prevalence of low level leaks.
We present the current status of knowledge regarding the potential environmental impacts of hydrogen releases and a high-level overview of projects funded by the U.S. Department of Energy's Hydrogen and Fuel Cell Technologies Office. Activities include collaboration with NOAA to improve understanding of atmospheric effects of hydrogen releases, and RD&D projects developing low level detection and quantification technologies, data collection of releases and leak rates, and advancement of release detection in indoor and outdoor environments.

In addition, we have developed the Hydrogen Component Reliability Database (HyCRED) project to collect reliability data that can be used in quantitative risk assessment, reliability, safety studies, maintenance planning, and more. The database is being piloted with hydrogen fueling stations with long term plans to expand the taxonomy and data collection across the hydrogen value chain. This presents a unique opportunity for proactive industry collaboration to provide data and real-world experience that will help understand how hydrogen component reliability impacts infrastructure availability and safe deployment, and may have important implications for providing objective information about hydrogen release events.

R&D Safety Investigations for Underground Liquid Hydrogen Storage

CONFERENCE

Center Hydrogen Safety Conference

YEAR

2024

PROCEEDING

2024 Center for Hydrogen Safety Americas Conference

GROUP

General Submissions

SESSION

Session 11 - New Applications and Applied R&D

TIME

Thursday, May 23, 2024 - 11:40am to 12:00pm

Authors

Houssin, D. - Presenter

According to preliminary studies, in addition to Hydrogen Refueling Stations layout flexibility, an underground hydrogen storage could bring at least 25% footprint reduction as compared to an above ground storage. Knowing that space is particularly a constraint in dense urban areas where hydrogen demand is high, underground configuration for liquid hydrogen storages could be an interesting alternative if the safety challenges induced by such a configuration are exhaustively inventoried and smartly addressed.
In this way, R&D led investigations in order to examine the different possible options of underground storage, define the most relevant configuration considering safety, technical and market constraints, and assess the feasibility to operate the selected configuration in safe conditions.

A panel of approaches was used - combining analytical simulations, lab-scale preliminary research, and near real-scale experimental simulations - to assess consequences of feared events, enrich risk analysis and propose concrete recommendations for safe deployment of the solution.

Thermal Radiation Investigations of Real-Scale Hydrogen Jet Flames at High Pressure

CONFERENCE

Center Hydrogen Safety Conference

YEAR

2024

PROCEEDING

2024 Center for Hydrogen Safety Americas Conference

GROUP

General Submissions

SESSION

Session 11 - New Applications and Applied R&D

TIME

Thursday, May 23, 2024 - 12:00pm to 12:15pm

Authors

Bernardy, C. - Presenter
Habib, K.
Orchini, A.

In order to reduce global warming, the use of hydrogen as a renewable energy source is becoming more important. To enable this transition, unprecedently large amounts of hydrogen need to be safely transported and stored. Since hydrogen is usually stored and transported under pressure, one scenario to be considered is the release of hydrogen from a leakage with subsequent ignition. The resulting jet flame must be characterized with respect to the thermal radiation emitted into the environment to define safety distances. Various models that characterize the resulting flame shape and radiation already exist in the literature, but these are mainly based on empirical data from hydrocarbon jet flames. To verify the applicability of these models to hydrogen, real-scale tests are carried out at the BAM Test Site Technical Safety (BAM-TTS) with the aim to assess the flame geometry and the emitted thermal radiation. Parameters such as leakage diameter (currently 1 mm to 10 mm), pressure (currently up to max. 250 bar) and mass flow (up to max. 0.5 kg/s) are varied. In particular, the focus is laid on the measurement and modelling of the thermal radiation. The challenge here is the characterization of the flame geometry in an open environment and its impact on the thermal radiation. Existing heat radiation data from the literature are mostly based on unsteady outflow conditions. The experimental setup used here allows for the generation of a steady-state outflow for several minutes and thus a direct comparability with existing (steady-state) models.

Preliminary Findings from Liquid Hydrogen Pooling and Vaporization Experiments

CONFERENCE

Center Hydrogen Safety Conference

YEAR

2024

PROCEEDING

2024 Center for Hydrogen Safety Americas Conference

GROUP

General Submissions

SESSION

Session 11 - New Applications and Applied R&D

TIME

Thursday, May 23, 2024 - 12:15pm to 12:30pm

Authors

Hecht, E. - Presenter, Sandia National Laboratories

We present preliminary findings from liquid hydrogen pooling and vaporization experiments performed in a controlled, constant cross-wind. The experiments investigated the effect of cross-wind flow-rate (from 1-5 mph), substrate type (steel and concrete), and liquid hydrogen spill rate (from 10-50 lpm) on the rate of pooling and downwind dispersion of liquid hydrogen. While previous experimental efforts worldwide have produced some data for liquid hydrogen pooling and vaporization, these experiments were unique in that they were performed in a semi-protected facility where the cross-wind flow rate was uni-directional, steady, and controlled, rather than relying on favorable ambient conditions. This experimental configuration produces data that is more suitable for both pooling and dispersion model validation than has been previously generated. In this presentation, the preliminary findings from these experiments are discussed as well as the potential implications for revising and developing science-based regulations, codes and standards for liquid hydrogen.

SNL is managed and operated by NTESS under DOE NNSA contract DE-NA0003525.

Pressure Relief Device Failure Taxonomy for a Proper Data Collection Strategy

CONFERENCE

Center Hydrogen Safety Conference

YEAR

2024

PROCEEDING

2024 Center for Hydrogen Safety Americas Conference

GROUP

General Submissions

SESSION

Session 12 - Safety Strategies

TIME

Thursday, May 23, 2024 - 2:00pm to 2:15pm

Authors

Jimenez, A. - Presenter, Engineering Systems Inc.
Groth, K., University of Maryland

Pressure relief devices (PRDs) are important safety devices that protect hydrogen equipment against excessive pressure increases that can damage the equipment. Typically, these PRDs are intended to mitigate overpressures caused by external fires or failures of upstream equipment. However, recent incidents indicate that PRDs are also a source of unintended hydrogen releases that can lead to fires or explosions. Furthermore, spurious opening failures can also lead to large hydrogen releases and undesired events such as a fire or an explosion. Given these new hazard characteristics, there is a need to better understand the risk-benefit tradeoff present when deciding to install PRDs in hydrogen systems; including under what conditions these devices provide a benefit. Quantitative risk analysis (QRA) provides a framework to assess the risk trade-off between the PRD capabilities to mitigate versus the times they produce an incident.

An important aspect of conducting this QRA is to establish a clear taxonomy of failure modes, failure mechanisms, and incident data to support this QRA. We reviewed hydrogen incidents involving PRDs, and used this review along with reliability engineering approaches to define proper PRD reliability taxonomies that can be used in a data collection and QRA. This presentation will summarize some of the PRD incidents identified during this review and will present the taxonomy and failure terminology identified for hydrogen PRD failure modes, operations, and incident types.

A Global Mapping of Best Practice Design Standards for Hydrogen Project Developers

CONFERENCE

Center Hydrogen Safety Conference

YEAR

2024

PROCEEDING

2024 Center for Hydrogen Safety Americas Conference

GROUP

General Submissions

SESSION

Session 12 - Safety Strategies

TIME

Thursday, May 23, 2024 - 2:15pm to 2:30pm

Authors

Sharpe, W. - Presenter
Crowley, C. - Presenter
Christoffersen, C. - Presenter

In a world that seeks cleaner energy sources, the emerging hydrogen sector remains one Hindenburg Disaster away from losing its social license and the current governmental support it needs for further development. Over the last decade, much progress has been made in relation to the standards that are required for the world to produce, store, transport and distribute hydrogen across the range of differing regulatory regimes and industrial uses. Unfortunately, gaps remain, and differences exist between what is presented as best practice across standards.

The objective of this paper is to allow the reader to understand the gaps and differences between standards in greater detail, whilst presenting a reference study for hydrogen project developers to use as they progress from concept to operations. Where standards lack guidance on project distinctions (e.g., novel pipeline configurations or ambitious operating capacity) this paper will outline what reference resources are available and what global project experience can teach the sector, to ensure a safe design is developed.

The paper will outline the output from a review and mapping of the current available international and national standards. The mapping will show the reader the gaps in coverage and provide insights in standardization approach used, the degree of prescription of use and will apply objective comparison and scoping criteria to inform what can be offered as best practice approach. Standard metadata will be compiled, including the level of prescription, adoption within each jurisdiction, year of issue and planned revisions, with cross-referencing to other standards where appropriate.

Ensuring Electrolyzer Safety: Key Strategies for Decarbonizing Industrial-Scale Hydrogen Production

CONFERENCE

Center Hydrogen Safety Conference

YEAR

2024

PROCEEDING

2024 Center for Hydrogen Safety Americas Conference

GROUP

General Submissions

SESSION

Session 12 - Safety Strategies

TIME

Thursday, May 23, 2024 - 2:30pm to 2:45pm

Authors

Schulthess, J. - Presenter
Panicker, J. - Presenter
Buchman, M. - Presenter
Panchula, A. - Presenter

Hydrogen electrolysis projects are growing rapidly in both quantity and scale. Demonstration projects are yielding to industrial scale projects with hundreds of megawatts of capacity. While the associated decarbonization benefits are exciting, it is essential to prioritize the safety of operators, communities, and the environment.

Electric Hydrogen takes the stance that safety during this time is paramount and a collaborative effort. Ensuring safety at large electrolysis plants benefits the entire industry by fostering trust with all project stakeholders. Electric Hydrogen has developed a comprehensive safety approach during the development of its 100MW product and wants to expand the discussion of electrolyzer safety by sharing unique risks and mitigation strategies associated with industrial scale electrolysis.

AUTHOR INDEX

AIChE
120 Wall Street, FL 23
New York, NY 10005-4020

ISBN 979-8-3313-0081-4

9 798331 300814

CURRAN ASSOCIATES INC.
proceedings
.com